GEMINI

The Mysterious AI Upgrade

Matthew H. Larsen

Copyright

Table of Content

Introduction

Gemini: The Mysterious AI Upgrade marks a significant milestone in the world of artificial intelligence (AI), marking the arrival of a new era in multimodal language models.

Developed by Google DeepMind, Gemini stands as a testament to the constant chase of innovation and progress in the field of machine learning. With its enigmatic nature and unprecedented powers, Gemini has caught the mind of researchers, developers, and fans alike.

The origins of Gemini goes back to the reveal of Google's big project during the Google I/O keynote on May 10, 2023.

Positioned as the successor to PaLM 2, Gemini appeared as a beacon of progress, offering to push the limits of what was previously thought possible in natural language understanding and creation.

Google CEO Sundar Pichai's news sparked a spark of interest and expectation within the tech community, setting the stage for the unveiling of this mysterious AI upgrade.

Unlike its predecessors, Gemini marks a departure from traditional methods to language modeling. Leveraging state-of-the-art methods and approaches, Gemini transcends the limits of text-based training, adding multimodal inputs to improve its knowledge of human language and context.

This unique method underscores Google DeepMind's dedication to promoting innovation and driving technological evolution.

Gemini's journey from idea to realization has been shrouded in secret and speculation. Throughout its developmental stages, details surrounding Gemini's design, training methods, and powers remained closely kept secrets, fueling speculation and interest within the AI community. As whispers of Gemini's presence got louder, so too did the anticipation surrounding its final launch.

On December 6, 2023, the veil of secrecy was broken as Google publicly announced the coming of Gemini to the world. Gemini is not merely a single entity but rather a family of large language models, each tailored to meet the varied needs of users across various areas. From the powerful Gemini Ultra to the flexible Gemini Nano, each member of the Gemini family brings its own unique skills and powers to the table.

This flexibility ensures that Gemini is well-equipped to handle a wide range of tasks and challenges, cementing its place as a formidable force in the world of AI. The introduction of Gemini has far-reaching effects for the future of AI and its uses. From conversational agents and virtual helpers to content generation and language translation, Gemini has the ability to change how we connect with technology and each other.

Background of Gemini

The inception of Gemini marks a culmination of years of study, development, and creation in the field of artificial intelligence. Rooted in Google DeepMind's relentless goal of pushing the limits of machine learning, Gemini appeared as a witness to the ever-evolving environment of AI technology.

The groundwork for Gemini was made with the study of large language models (LLMs), which have changed natural language processing (NLP) in recent years. Building upon the success of models like BERT and GPT, Google sought to build a next-generation LLM that could not only understand and produce text but also combine multimodal inputs for a more complete understanding of language and context.

The need for such a model became increasingly clear as AI uses spread beyond simple text-based interactions. From image recognition to voice orders, current AI systems are tasked with processing a wide array of inputs to provide useful answers and insights.

Gemini aimed to handle this issue by integrating multiple modalities, including text, images, and sounds, into its language understanding system.

The quest for a multimodal LLM led Google DeepMind to study novel architectures and training methods. Drawing inspiration from recent gains in deep learning and neural network research, the team went on a journey to create a model that could smoothly combine information from diverse sources while maintaining efficiency and scale.

One of the key breakthroughs in Gemini's growth was its departure from standard text-based training methods. Instead of depending solely on huge text corpora, Gemini leveraged multimodal datasets holding a varied range of inputs, including text, pictures, and audio.

This multimodal method helped Gemini to learn from a more comprehensive set of facts, allowing it to better understand and analyse human language in context.

Another important aspect of Gemini's background lies in its predecessors, especially LaMDA and PaLM 2. LaMDA, short for Language Model for Dialogue Applications, focused on conversational understanding and creation, setting the groundwork for Gemini's chatbot powers.

PaLM 2, or Probabilistic and Logical Modeling, brought advancements in probabilistic reasoning and logical inference, which were further refined and added into Gemini's design.

The creation of Gemini was characterized by a collaborative effort involving researchers, engineers, and subject experts from across Google DeepMind and its parent company, Alphabet.

Drawing upon diverse knowledge and views, the team iteratively refined Gemini's design and usefulness, constantly pushing the edges of what was achievable with AI technology.

Throughout its development, Gemini stayed shrouded in secret, with only limited information shared with the public. This air of mystery surrounding Gemini only served to heighten expectations and rumour within the AI community, fueling excitement for its final unveiling.

On May 10, 2023, the curtain of secret was partly broken as Google publicly announced the presence of Gemini during the Google I/O keynote. While details remained scarce, the news created considerable buzz and speculation, marking the dawn of a new age in AI research and development.

In the months that followed, Google DeepMind continued to improve and optimize Gemini, harnessing the power of advanced computational tools and cutting-edge algorithms. On December 6, 2023, Gemini was finally revealed to the world, marking a major milestone in the evolution of AI technology.

Development of Gemini

The development journey of Gemini represents a collaborative effort driven by innovation, commitment, and a relentless chase of greatness within Google DeepMind. From its conceptualization to its ultimate reality as a groundbreaking multimodal language model, Gemini's development unfolded through a series of iterative improvements and breakthroughs in AI research and engineering.

The origin of Gemini can be traced back to the ongoing quest within Google DeepMind to push the limits of natural language understanding and generation. Building upon the wins and lessons learned from earlier projects, including LaMDA and

PaLM 2, the team went on a mission to build a next-generation AI model capable of smoothly combining multiple forms of input.

Early in the development process, the team identified several key goals that would shape Gemini's design and functionality. Chief among these goals was the need to move beyond standard text-based training methods and add multimodal datasets to improve the model's knowledge of language in context.

To achieve this goal, Google DeepMind applied state-of-the-art methods and approaches from the areas of computer vision, audio processing, and natural language processing. By combining ideas from these diverse areas, the team tried to build a model that could successfully process and understand information from a wide range of sources, including text, images, and audio. One of the primary challenges met during Gemini's development was creating a system capable of quickly handling multimodal inputs while maintaining scalability and processing efficiency.

Traditional neural network designs were ill-suited for handling multimodal data, often leading to speed problems and higher computational overhead. To handle this issue, the team studied novel architectural forms inspired by recent improvements in deep learning research. This led to the development of a custom design made especially for Gemini, capable of smoothly fusing information from different channels while optimizing computing resources.

In parallel with architectural design, significant efforts were spent in improving training methods and algorithms to ensure efficient and effective learning from multimodal datasets. This involved the development of specialized training methods suited to the unique features of Gemini's design, such as attention mechanisms and multi-task learning strategies.

Throughout the development process, Google DeepMind kept a culture of collaboration and interdisciplinary study, drawing upon the knowledge of researchers, engineers, and domain experts from across the company. This joint approach allowed fast iteration and testing, allowing the team to quickly improve on ideas and incorporate feedback into the development process.

As Gemini moved through its developmental stages, it faced rigorous testing and review to prove its performance and capabilities. Benchmarking against current models and real-world use cases offered useful insights into Gemini's strengths and areas for improvement, guiding further refinement and optimization efforts.

The culmination of years of study, development, and iteration came to fruition on December 6, 2023, when Google publicly announced the start of Gemini to the world. Positioned as a competitor to OpenAI's

GPT-4, Gemini marked a major leap forward in the field of multimodal language modeling, offering unique powers and performance.

In the years since its start, Gemini has continued to grow and improve, driven by ongoing research and development efforts within Google DeepMind and the broader AI community. With each iteration, Gemini pushes the edges of what is possible in natural language understanding and generation, paving the way for new and interesting uses across a wide range of fields. The creation of Gemini acts as a testament to the power of cooperation, creativity, and perseverance in advancing the frontiers of artificial intelligence.

Announcement and Release

The news and release of Gemini marked a pivotal moment in the landscape of artificial intelligence, catching the attention of researchers, developers, and fans worldwide. It marked the result of years of research and development within Google DeepMind, marking a major leap forward in the field of multimodal language models.

The first whispers of Gemini appeared during the Google I/O keynote on May 10, 2023, when Google CEO Sundar Pichai revealed the project to the world. Positioned as the successor to PaLM 2 and LaMDA, Gemini claimed to change natural language understanding and generation with its novel approach to multimodal learning.

Pichai's statement sent shockwaves through the AI community, sparking speculation and expectation surrounding the powers and consequences of Gemini. While details remained scarce at the time, the mere mention of Google's latest AI attempt sparked a flurry of excitement and curiosity among researchers and fans alike.

In the months that followed, Google DeepMind stayed relatively tight-lipped about Gemini, keeping information surrounding its development and features under wraps. This air of mystery only served to heighten anticipation as the tech world eagerly watched further news and reports on the project.

Finally, on December 6, 2023, the veil of secrecy was broken as Google publicly announced the release of Gemini to the public. Positioned as a contender to OpenAI's GPT-4, Gemini made its big entry onto the AI stage, captivating viewers with its promise of unparalleled performance and flexibility.

The announcement of Gemini was met with broad joy and enthusiasm, as researchers and developers clamored to get their hands on the latest improvements in AI technology. From industry professionals to hobbyists, the release of Gemini marked a chance to discover new options and push the boundaries of what was possible with AI.

One of the key points of Gemini's release was its family of models, which catered to a diverse range of use cases and uses. From the powerful Gemini Ultra to the lightweight Gemini Nano, each member of the Gemini family brought its own unique strengths and powers to the table, ensuring that there was a model fit for every need and budget.

In the weeks and months following its release, Gemini gained broad attention and praise for its impressive performance and versatility. Researchers and developers quickly began adding Gemini into their projects, exploring its powers across a wide range of fields, from conversational agents and virtual helpers to content generation and language translation.

Despite its initial success, Gemini met its fair share of challenges and complaints, especially regarding issues of openness, bias, and ethical considerations. As with any new technology, the release of Gemini sparked important talks within the AI community about the responsible development and usage of AI systems.

Moving forward, Google DeepMind stays dedicated to improving the powers of Gemini and handling any concerns or complaints that may emerge. With ongoing research and development efforts, Gemini continues to grow and improve, paving the way for new and exciting uses in the field of artificial intelligence.

The announcement and release of Gemini marked a new chapter in the story of AI, one defined by innovation, cooperation, and the relentless chase of progress. As researchers and producers continue to study the prospects of Gemini, the future of artificial intelligence looks brighter than ever before.

Gemini Family

Gemini Ultra

Gemini Ultra stands as the peak of creativity within the Gemini family, reflecting the cutting edge of multimodal language modeling technology. Engineered for unmatched performance and flexibility, Gemini Ultra redefines the limits of what is possible in natural language understanding and creation.

At the heart of Gemini Ultra lies a sophisticated architecture intended to smoothly combine multiple forms of input, including text, images, and voice. This multimodal approach allows Gemini Ultra to glean deeper insights from diverse sources of information, resulting in more nuanced and contextually rich knowledge of human language.

One of the key distinguishing features of Gemini Ultra is its huge size and computational power. Equipped with an expansive neural network design and access to vast computing resources, Gemini Ultra is capable of processing and analyzing immense amounts of data with lightning-fast speed and accuracy.

The sheer scale of Gemini Ultra allows it to handle complicated tasks and problems that were previously thought to be beyond the capabilities of AI. From conversational bots and virtual helpers to content generation and language translation, Gemini Ultra excels in a wide range of applications, providing human-like answers and insights with unparalleled accuracy and fluency.

But Gemini Ultra is more than just a powerful AI model—it's also a platform for creation and discovery. Researchers and developers around the world are harnessing the capabilities of Gemini Ultra to push the limits of AI technology, discovering new use cases and uses that were previously unthinkable.

Despite its amazing powers, Gemini Ultra is not without its challenges and limits. Issues such as model bias, ethical issues, and computational tools pose ongoing challenges for academics and producers working with Gemini Ultra. However, with ongoing research and development efforts, Google

DeepMind stays dedicated to addressing these challenges and unlocking the full potential of Gemini Ultra. As the top model within the Gemini family, Gemini Ultra marks the peak of success in AI research and development. With its unmatched speed, flexibility, and potential for innovation, Gemini Ultra is set to shape the future of artificial intelligence and change how we interact with technology and each other.

Gemini Pro

Gemini Pro appears as a versatile and adaptable member of the Gemini family, built to cater to a wide range of use cases and applications. Positioned between the flagship Gemini Ultra and the lightweight

Gemini Nano, Gemini Pro strikes a balance between speed and economy, making it an ideal choice for various AI-driven tasks.

While not as powerful as Gemini Ultra, Gemini Pro boasts amazing features that set it apart from other big language models on the market. Equipped with a refined design optimized for efficiency and scaling, Gemini Pro provides exceptional performance across a diverse range of applications, from conversational bots to content generation.

One of the key strengths of Gemini Pro lies in its adaptability to different computational settings and resource limits. Whether installed on cloud infrastructure or integrated within edge devices, Gemini Pro uses its streamlined design to maximize performance while reducing resource usage, making it suitable for deployment in a variety of situations.

Despite its more modest computing needs, Gemini Pro does not compromise on quality or accuracy. Leveraging advanced training techniques and algorithms, Gemini Pro achieves amazing levels of fluency and clarity in its created outputs, ensuring a seamless user experience across a wide range of apps.

With its flexibility and adaptability, Gemini Pro serves as a flexible tool for researchers, developers, and groups looking to harness the power of AI in their projects and products. Whether used for making conversational interfaces, generating creative content, or analyzing big amounts of data, Gemini Pro offers a flexible and scalable option for a variety of use cases.

As the middle child of the Gemini family, Gemini Pro represents a balanced approach to AI modeling, offering an appealing mix of speed, efficiency, and flexibility. With its ability to adapt to different computing settings and applications, Gemini Pro stands as a proof to Google DeepMind's commitment to providing new solutions that meet the varied needs of its users.

Gemini Nano

Gemini Nano appears as the lightweight and efficient member of the Gemini family, built to bring the power of AI to resource-constrained environments and edge devices. With its compact design and minimal computational needs, Gemini Nano opens up new possibilities for AI-driven applications in areas where standard models may not be possible.

Despite its smaller size, Gemini Nano maintains many of the core capabilities and features of its larger versions, including advanced natural language understanding and generation capabilities. Leveraging new methods and algorithms, Gemini Nano achieves amazing levels of speed and accuracy, making it perfect for a wide range of uses.

One of the key benefits of Gemini Nano lies in its ability to run quickly on devices with limited computational resources, such as smartphones, IoT devices, and embedded systems. By leveraging its compact design and optimized algorithms, Gemini Nano provides AI-powered experiences directly at the edge, without the need for constant access or cloud infrastructure.

The lightweight nature of Gemini Nano also makes it an ideal choice for privacy-sensitive apps where data must be handled locally without being transmitted to external servers. By keeping data on-device, Gemini Nano allows users to maintain greater control over their personal information while still gaining from the power of AI.

Despite its compact size, Gemini Nano does not skimp on quality or efficiency. Whether used for building intelligent IoT devices, enhancing mobile apps, or allowing edge computing scenarios, Gemini Nano provides robust and reliable AI capabilities that empower developers to create new and impactful solutions.

As the smallest member of the Gemini family, Gemini Nano marks a major step forward in the democratization of AI, bringing the power of intelligent computing to the edge of the network. With its compact design, minimal computational needs, and amazing performance, Gemini Nano opens up new possibilities for AI-driven innovation in a wide range of fields.

Features and Capabilities

Gemini, the new multimodal language model created by Google DeepMind, offers a wide array of features and powers that set it apart from standard big language models. From its original design to its strong generative capabilities, Gemini offers a wealth of functionality that allows it to excel in a variety of AI-driven applications.

1. Multimodal Learning:

Gemini's most distinguishing trait is its ability to handle and understand multiple types of information, including text, images, and voice. Unlike traditional big language models that depend solely on text-based data, Gemini uses multimodal datasets to achieve a deeper and more complete knowledge of human language and context.

2. Versatility and Adaptability:

Gemini's family of models, which includes Gemini Ultra, Gemini Pro, and Gemini Nano, gives a range of choices to fit different computing settings and use cases. Whether deployed in the cloud, on edge devices, or embedded within apps, Gemini adapts to different resource limits while keeping high levels of speed and efficiency.

3. High-speed Computing:

Gemini Ultra, the top model within the Gemini family, boasts unparalleled computational power and speed. Equipped with an expansive neural network design and access to vast computer resources,

Gemini Ultra excels in jobs requiring high-performance computing, such as natural language understanding, creation, and reasoning.

4. Efficient Edge Computing:

Gemini Nano, the lightweight member of the Gemini family, is designed for edge computing situations where processing resources are limited. Despite its small size, Gemini Nano provides powerful AI capabilities directly at the edge, allowing intelligent processing and decision-making in real-time, without the need for constant access or cloud infrastructure.

5. Natural Language Understanding and Generation:

At its core, Gemini is meant to understand and create human-like answers and insights. Leveraging advanced techniques and algorithms, Gemini excels in jobs such as conversational agents, virtual helpers, content generation, and language translation, providing natural and smooth interactions across a wide range of applications.

6. Ethical Considerations and Bias Mitigation:

Gemini is provided with methods to reduce bias and ensure ethical AI practices. By directly addressing issues of justice, transparency, and responsibility, Google DeepMind aims to foster trust and faith in

Gemini's powers while minimizing the potential for harmful or unintended effects.

7. Developer Support and Ecosystem:

Google DeepMind offers extensive guidance, tools, and resources to support developers in building and launching AI-driven apps with Gemini. From model training and review to deployment and tracking, Google DeepMind's ecosystem enables developers to unlock the full potential of Gemini in their projects and products.

Overall, Gemini's features and powers place it as a versatile and powerful tool for AI creation and experimentation. Whether applied in high-performance computer settings or at the edge of the network,

Gemini offers a scalable and efficient solution for a wide range of AI-driven applications, paving the way for new possibilities in natural language understanding and generation.

Use in Generative Chatbot

Gemini's advanced capabilities make it an ideal choice for creating generative chatbots, allowing natural and engaging talks between users and AI-driven interfaces.

Leveraging its multimodal learning framework and advanced natural language understanding and generation powers, Gemini takes the conversational experience to new heights, providing human-like interactions and answers that feel real and intuitive.

1. Natural Language Understanding:
Gemini's ability to understand human language in context is important for making effective generative robots. By analyzing text inputs and pulling meaning from talks, Gemini can understand user goals, extract relevant information, and react correctly, ensuring a seamless and intuitive user experience.

2. Contextual Understanding:
Gemini excels at keeping context and coherence throughout talks, allowing chatbots to understand and react to user inputs in a useful way. Whether continuing a chat thread or handling pauses and changes

in topic, Gemini ensures that interactions feel natural and fluid, creating a sense of continuity and engagement.

3. Multimodal Inputs:

One of Gemini's key strengths comes in its ability to handle multimodal inputs, including text, pictures, and voice. This allows chatbots to connect with users in more dynamic and engaging ways, incorporating visual and auditory cues into talks for a deeper and more immersive experience.

4. Creative Content Generation:

Gemini's generative powers stretch beyond simple answers to user questions, allowing chatbots to create creative and engaging content on the fly. From creating personalized suggestions to writing engaging tales, Gemini equips robots to create content that captivates and delights users, improving the overall conversational experience.

5. Personalization and Adaptation:
Gemini's ability to learn and change over time allows robots to adjust conversations based on user preferences, behaviors, and feedback. By studying past exchanges and user data, Gemini can tailor answers and suggestions to each individual user, creating a more personalized and engaging conversational experience.

6. Ethical Considerations:
Google DeepMind puts a strong focus on ethical AI practices, ensuring that

Gemini-powered chatbots stick to principles of justice, transparency, and responsibility. By directly handling issues of bias, privacy, and consent, Gemini-powered chatbots promote trust and confidence among users, promoting positive interactions and relationships.

Gemini's use in generative chatbots marks a major advancement in the field of conversational AI, giving a powerful and versatile platform for building intelligent interfaces that engage and please users. With its advanced natural language understanding and generation capabilities, Gemini-powered chatbots are set to change the way we interact with AI-driven interfaces, bringing in a new era of personalized and interactive conversational experiences.

Comparison with Predecessors (LaMDA and PaLM 2)

Gemini marks the latest evolution in Google DeepMind's goal of advancing natural language processing (NLP) skills. To understand its significance, it's necessary to compare and contrast Gemini with its predecessors, LaMDA and PaLM 2, each adding unique strengths and innovations to the field of AI.

1. LaMDA (Language Model for Dialogue Applications): LaMDA worked on improving conversational understanding and generation, hoping to build chatbots capable of participating in more natural and

contextually rich talks. Unlike standard language models that produce responses based solely on textual patterns, LaMDA hoped to understand the underlying context of a conversation, allowing more coherent and meaningful exchanges.

In comparison, Gemini builds upon LaMDA's base by adding multimodal learning capabilities, allowing it to understand and process multiple types of input, including text, pictures, and voice. While LaMDA excelled in conversational settings, Gemini's multimodal method expands its application to a wider range of tasks and use cases, making it more flexible and adjustable.

2. PaLM 2 (Probabilistic and Logical Modeling): PaLM 2 brought advancements in probabilistic reasoning and logical inference, allowing AI models to reason and make choices based on uncertain or incomplete information.

By combining probabilistic and logical methods, PaLM 2 aims to improve the stability and trustworthiness of AI systems, particularly in complex and ambiguous settings.

Gemini builds upon PaLM 2's foundational concepts by integrating probabilistic reasoning and logical inference skills into its design. This helps Gemini to make more informed choices and generate responses based on a better knowledge of context and uncertainty. By leveraging these capabilities, Gemini improves the quality and accuracy of its exchanges, making it more effective in real-world applications.

while LaMDA and PaLM 2 laid the groundwork for advancements in conversational understanding and logical reasoning, respectively, Gemini marks a major leap forward by combining these strengths with multimodal learning capabilities.

By integrating text, pictures, and voice inputs, Gemini gives a more complete and nuanced understanding of human language and context, paving the way for more intelligent and adaptable AI systems.

Licensing and Availability

Gemini, created by Google DeepMind, is made available to users and developers under a private license. This licensing model provides Google DeepMind with power over the spread, usage, and modification of Gemini, ensuring that it stays compliant with legal and ethical standards while also protecting intellectual property rights.

As a private software product, Gemini is subject to terms and conditions stated by Google DeepMind, which control its usage and distribution. These terms may include limitations on business use, redistribution, and change, as well as requirements for credit and compliance with relevant laws and regulations.

Gemini's access to users and coders may change based on their location, connection, and planned use case. Google DeepMind may give access to Gemini through different channels, including study collaborations, academic partnerships, and business licensing deals.

For researchers and coders interested in studying Gemini's powers, Google DeepMind may provide access to pre-trained models, documentation, and tools through its website or development portals. These tools allow users to play with Gemini, build applications, and integrate it into their projects.

Commercial licensing agreements may be offered for groups looking to apply Gemini in commercial goods or services. These agreements usually involve negotiations with Google DeepMind to determine licensing fees, usage rights, and support services tailored to the organization's unique needs and requirements.

In addition to licensing, Google DeepMind may also give support and maintenance services for Gemini, including software updates, bug fixes, and expert help. These services help ensure the continued dependability and performance of Gemini in live settings, providing users and developers with peace of mind as they leverage its capabilities.

Overall, while Gemini is offered under a private license, Google DeepMind tries to make it accessible to a wide range of users and developers through various channels and licensing choices. By giving access to documents, tools, and support services, Google DeepMind empowers users to harness the power of Gemini and drive innovation in the field of artificial intelligence.

Conclusion

Gemini represents a new development in the field of artificial intelligence, giving a powerful and flexible tool for natural language understanding and creation. With its innovative bidirectional learning capabilities, advanced design, and extensive family of models, Gemini pushes the limits of what is possible in AI technology, paving the way for new possibilities in communication, teamwork, and creativity.

From its inception, Gemini has caught the mind of researchers, developers, and fans worldwide, sparking excitement and anticipation for its potential to revolutionize how we communicate with technology and each other. Its ability to process and understand multiple forms of input, including text, images, and voice, allows Gemini to achieve a deeper and more detailed knowledge of human language and context, creating more natural and engaging interactions.

Gemini's family of models, including Gemini Ultra, Gemini Pro, and Gemini Nano, offers choices to fit a wide range of computational settings and use cases, from high-performance computing to edge computing scenarios. This flexibility ensures that Gemini is available to users and developers across diverse fields, enabling them to leverage its powers in their projects and products.

Despite its impressive capabilities, Gemini is not without its difficulties and considerations, including problems of bias, ethics, and privacy. Google DeepMind stays dedicated to addressing these issues through ongoing study, development, and collaboration with the AI community, ensuring that Gemini remains at the forefront of responsible and ethical AI innovation.

As we look to the future, the promise of Gemini is endless. Whether generating generative chatbots, allowing intelligent helpers, or supporting creative content generation, Gemini opens up new possibilities for innovation and discovery in AI-driven apps. With its continued development and breakthroughs, Gemini promises to shape the future of artificial intelligence, opening new possibilities and changing how we interact with technology in the years to come.